Spike

A Story of Courage and Love

True Tales from
Emerald Springs Farm
Polk County, North Carolina

Written and Illustrated by Lee Barker

This book is dedicated to:

Bill, our farmer, who with hard work and love has created a chicken, turkey, goat, dog, cat, parakeet, fish and human-friendly farm,

and

Jessie, with thanks for helping my eyes to see our animal partners with respect and compassion.

Spike - A Story of Courage and Love

Book I in the Series

True Tales from Emerald Springs Farm

located in beautiful

Polk County, North Carolina

written and illustrated by

Lee Barker

Emerald Springs (almost) Free Press

A portion of the proceeds from this book will be donated to the Foothills Humane Society of Polk County, NC

Emerald Springs Farm

Nine Acres of Heaven in the Foothills of Polk County, North Carolina

One of good mama's brood was a hen named Spike.
She seemed to be just another hen;

foraging in the sunlit pastures

roaming through the cool forest

adventuring in the bamboo jungle...

hiding from the hawks and running from the foxes.

It took a while for the farmer to see that Spike was special.

At the end of her first year, Spike went broody –
it was time to sit on eggs!

She fiercely defended her nest, growling and pecking anyone who came near.

Still the farmer managed to steal her egg every day.

One day, the farmer locked Spike in a small crate near the barn. Inside there was food and water and a clean nest with 5 fresh eggs to sit on.

No one came to take them away and Spike faithfully kept them warm and clean for 3 long weeks.

When hatching day came, Spike watched with joy as each tiny beak broke through its hard brown shell. She waited patiently and clucked encouragement deep in her throat.

The farmer was watching too, and when the last egg hatched, the door of the crate opened.

Spike led her tiny brood out into the wide world. Every day she taught them how to hunt and scratch for food. Every day she protected them from danger –

Dogs!

Cats!

Other Chickens!

And Farmers!

Every night she and her brood returned to their cozy nest.

Then tragedy struck.

Spike and her chicks were hunting near the edge of the bamboo jungle when a wily fox rushed out of the darkness.

Before Spike could move, the fox snatched a chick and disappeared, leaving only a single feather floating to the ground.

It was a mother fox with two fine kits.

They romped and played in the kudzu ravine while she hunted for their food.

Now, Spike was a very smart chicken and Spike remembered it all.

Through the glowing colors of Fall.

Through the snows of a cold Winter.

Into the bright promise of a new Spring.

That Spring at broody time, Spike quietly disappeared. Other hens hatched out their broods in the farmer's nesting boxes and the pastures filled with small chicks and mama hens. Hawks and foxes came often, but most chicks survived and grew.

One day near the end of broody time, the farmer was feeding some delicious scratch to the whole clan when Spike stepped out of the bushes.

Following her comforting clucks were eleven tiny chicks! Spike had made a secret nest and filled it with her very own eggs! What a fine, clever mama hen Spike was!

But she did even better.

Unlike the other hens, Spike never led her brood down to the bamboo jungle where the foxes lurked.

Instead she taught them to hunt and scratch in the bushes and gardens around the farmer's house. None of Spike's eleven chicks were lost to the foxes!

And Spike had even more surprises in store for the farmer.

Spike and her brood were making their rounds near the house when a large hawk spotted them !

Muffy, the livestock guard dog, was asleep. She'd been up all night barking away dogs, coyotes and foxes.

The farmer was still in the house eating breakfast.

No one was there to defend the half grown chicks from the hawk.   No one but Spike!

The hawk swooped down near a quivering chick. Spike shrieked and every feather stood on end. She screamed again and attacked!

Inside, the farmer heard Spike's fierce cries and rushed to the door.

And the farmer saw ...

The hawk, taking off empty clawed...

and Spike rushing to her chick !

Spike and her chicks hurried to the shadow of their favorite bushes.

A 4 foot long black snake slid by on its way to the pasture. Another predator who had learned to stay away from Spike's chicks!

“How does she do it?” puzzled the farmer.

She does it by being strong and brave and smart. Best of all, she does it because Spike is a VERY good mother!

Spike is a real chicken who really did all the things in this story. She and her clan are free ranging Golden Laced Wyandottes – large healthy chickens who find most of their own food all year round and who lay beautiful brown eggs. These chickens have their own language with a big vocabulary. They have 'friends' they like to be with and really love exploring and foraging on the nine acres here at Emerald Springs Farm.

Spike is a special chicken in many ways and has stayed with her chicks much longer than any of the other hens.

Made in the USA
Las Vegas, NV
10 November 2026

57851002R00017